Chapter 6 Resource File
with Answer Key

Holt Social Studies

World Geography

HOLT, RINEHART AND WINSTON

A Harcourt Education Company

Orlando • **Austin** • New York • San Diego • Toronto • London

Canada   Vocabulary Builder

Section 1

Canadian Shield	Grand Banks	Hudson Bay
newsprint	Niagara Falls	pulp
Rocky Mountains	St. Lawrence River	tundra
uplands		

DIRECTIONS Read each sentence and fill in the blank with the word in the word pair that best completes the sentence.

1. The _______________________ forms a natural border in the east between the United States and Canada. (**Canadian Shield/St. Lawrence River**)

2. _______________________ refers to softened wood fibers used to make paper. (**Pulp/Newsprint**)

3. The _______________________ extend from the western United States into western Canada. (**Grand Banks/Rocky Mountains**)

4. The Canadian Shield is a region of lakes, swamps, and rocky

    _______________________. (**tundra/uplands**)

5. _______________________ is a physical feature that the United States and Canada share. (**Hudson Bay/Niagara Falls**)

DIRECTIONS Choose five of the terms from the word bank. Use these terms to write a summary of what you learned in the section.

____

____

____

____

____

Canada

Vocabulary Builder

Section 2

DIRECTIONS On the line provided before each statement, write **T** if a
statement is true and **F** if a statement is false. If the statement is false,
write the term that would make the statement correct on the line after
each sentence.

______ **1.** Since World War II, <u>Quebec</u> has become one of the most culturally diverse
cities in the world.

____

______ **2.** Quebec and Ontario are examples of Canadian <u>colonies</u>.

____

______ **3.** The Canadians built a transcontinental railroad to connect <u>British
Columbia</u> with eastern Canada.

____

______ **4.** <u>Toronto</u> remains a mainly French-speaking region.

____

______ **5.** A <u>dominion</u>, such as the Dominion of Canada, is a territory or area of
influence.

____

DIRECTIONS Write three adjectives or descriptive phrases that describe
the term.

6. British Columbia ____

7. New France ____

8. Canadian Pacific Railroad ___________________________________

9. Vancouver's Chinatown _______________________________________

10. First Nations ____

Canada

Vocabulary Builder

Section 3

heartland	Inuit	maritime
Montreal	motto	Ottawa
regionalism	Toronto	Vancouver

DIRECTIONS Look at each set of four vocabulary terms. On the line provided, write the letter of the term that does not relate to the others.

______ **1. a.** culture
      b. regionalism
      c. connection
      d. industrial

______ **2. a.** coast
      b. heartland
      c. Atlantic
      d. maritime

______ **3. a.** Montreal
      b. Toronto
      c. Ottawa
      d. Windsor

______ **4. a.** Yukon
      b. Vancouver
      c. Nunavut
      d. tundra

______ **5. a.** Alberta
      b. Manitoba
      c. Saskatchewan
      d. Nova Scotia

DIRECTIONS Choose five terms from the word bank. On the lines below use these terms to write a poem or story that relates to the section.

____

____

____

____

____

____

____

Canada   Biography

Douglas Jung
1925–2002

HOW HE AFFECTED THE REGION Douglas Jung is known for his efforts to improve the lives of Chinese-Canadians. Born in a time of fierce racial discrimination, he joined the Canadian Army. Later he earned a law degree and became the first Chinese-Canadian Member of Parliament. Jung will be remembered for the differences he made in both Asian and Canadian cultures.

As you read the biography below, think about how Jung's pride in both his Chinese descent and his Canadian citizenship led him to do great things for both cultures.

Given his roots, Douglas Jung was unlikely to become one of the most honored Canadians of his time. Jung was born in Vancouver, British Columbia, in February 1925. At the time, Canada was not a welcoming place to Asian immigrants. They did not enjoy the same rights as non-Asian Canadians. But Jung hoped to create better conditions for everyone in Canada.

There was widespread **prejudice** against Chinese and other Asians in Canada. For example, Asians in Canada could not join the army. Only when World War II broke out did this change. Jung **enlisted** soon after it was legal to do so. He was one of the first Chinese-Canadians in the army. During World War II, Jung was assigned to a secret mission. He was sent to Malaysia, a place of fierce fighting, in Southeast Asia. There, his unit battled to free prisoners of war. Jung and other Asian-Canadian soldiers also fought a different battle. They fought to prove themselves the equal of other Canadian soldiers.

VOCABULARY

prejudice judging someone unfairly; discrimination

enlisted joined the army or another branch of the armed services

Parliament the lawmaking body of Canada's government

After the war, Jung was one of the first Chinese-Canadians to earn a law degree. Later, he opened his own law office. In 1957, Jung made history again. He was elected to the Canadian **Parliament**. Jung became its first Chinese-Canadian member ever. During his five-year term, he also represented Canada at the United Nations.

Jung became a judge in 1962. That year, he introduced a program that would change Canada. It would help thousands of Asian-Canadians to become legal citizens. Until his death in 2002, Jung kept fighting for the rights of all Canadian people.

WHAT DID YOU LEARN?

1. Recall How was Douglas Jung's military service an achievement for the Chinese-Canadian community?

____

____

2. Expressing and Supporting a Point of View What do you think was Douglas Jung's greatest accomplishment? Provide reasons or examples to support your point of view.

____

____

ACTIVITY

Imagine that you are a producer for a television program that tells the life stories of famous people. You are assigned a segment on Douglas Jung. Research key people who lived in Canada during Jung's time. Decide who you would want to interview for the segment. Explain how each one could contribute to Jung's story.

Canada   Biography

Mary Pickford
1892–1979

HOW SHE AFFECTED THE REGION Of the handful of actors and filmmakers who started the Hollywood film industry, many were Canadian. Born and raised in Toronto, Mary Pickford become one of the first big movie stars. She was also a co-founder of a major film studio. Her life and career remain a source of great pride for her native country.

As you read the biography below, think about how Pickford's deternimation earned her a place in both Canadian and Hollywood history.

They called her "America's Sweetheart." But Toronto, Canada, is where Mary Pickford's life and career began.

Born Gladys Smith in 1892 in Toronto, Pickford was one of three children. Her father died suddenly when she was young. To help make ends meet, Pickford started acting in plays in Toronto. She soon became a popular child star of the stage. Fans gave her the nickname "Baby Gladys."

Pickford had a strong desire to help her mother and **siblings**. As a teen, Pickford decided to head to New York City. She hoped to make it big on Broadway. "I was terribly ambitious you see. I was the father of the family," she once said. "I wanted security for my people."

Success came quickly for Pickford. She began working both on Broadway and in silent films. By age 23, she was Hollywood's most famous and highly-paid actress.

Acting wasn't Pickford's only strength. She was also a sound businesswoman. In 1919, she and actor Douglas Fairbanks teamed up with film legend

VOCABULARY
siblings brothers and sisters
founded started or created
legacy a reminder

Charlie Chaplin. Together, they **founded** the United Artists Corporation. Today, UA remains one of Hollywood's major film studios.

For her career, Mary Pickford earned an Oscar Award and a Lifetime Achievement Award. Still, she would often speak of her early life in Toronto. After Pickford's death, Canada awarded her a great honor. A star on Canada's Walk of Fame was dedicated to her. These awards serve as a **legacy** of her success on both sides of the border.

WHAT DID YOU LEARN?

1. Recall Why did Mary Pickford come to New York City?

____

____

2. Draw a Conclusion Why do you think Mary Pickford was known as "America's Sweetheart"?

____

____

ACTIVITY

Mary Pickford appeared in more than 50 feature films. Using the Internet or a library resource center, look for photos from some of these films. Print or copy them to create a collage of highlights from Pickford's career. Be sure to identify movie titles and years on your collage.

Canada   Literature

Anne of Green Gables
by Lucy Maud Montgomery

ABOUT THE READING *Anne of Green Gables* is the classic tale of Anne Shirley, a young orphan girl. Anne is sent to live with an elderly brother and sister on their farm on Prince Edward Island. Like Anne, the author grew up on the island, located off the coasts of Nova Scotia and New Brunswick. Since the novel's first publication in 1908, Anne Shirley has become a beloved character among young people worldwide.

VOCABULARY

bound leap

sash movable part of a window

boughs branches

thickset growing thickly

As you read the passage below, note the description of Green Gables, the name of the farm. Determine what this description might tell you about the story's larger setting, Prince Edward Island.

It was broad daylight when Anne awoke and sat up in bed, staring confusedly at the window through which a flood of cheery sunshine was pouring and outside of which something white and feathery waved across glimpses of blue sky.

For a moment she could not remember where she was. First came a delightful thrill, as of something very pleasant; then a horrible remembrance. This was Green Gables and they didn't want her because she wasn't a boy!

> Circle the phrases that tell how Anne feels when she wakes up.

But it was morning and, yes, it was a cherry-tree in full bloom outside of her window. With a **bound** she was out of bed and across the floor. She pushed up the **sash**—it went up stiffly and creakily, as if it hadn't been opened for a long time, which was the case; and it stuck so tight that nothing was needed to hold it up.

Source: From *Anne of Green Gables* by Lucy Maud Montgomery. New York, NY: Bantam Books, 1976.

Anne of Green Gables, *continued*   Literature

Anne dropped on her knees and gazed out into the June morning, her eyes glistening with delight. Oh, wasn't it beautiful? Wasn't it a lovely place? Suppose she wasn't really going to stay here! She would imagine she was. There was scope for imagination here.

A huge cherry-tree grew outside, so close that its **boughs** tapped against the house, and it was so **thickset** with blossoms that hardly a leaf was to be seen. On both sides of the house was a big orchard, one of apple trees and one of cherry trees, also showered over with blossoms; and their grass was all sprinkled with dandelions. In the garden below were lilac trees purple with flowers, and their dizzily sweet fragrance drifted up to the window on the morning wind . . .

Off to the left were the big barns and beyond them, away down over green, low-sloping fields, was a sparkling blue glimpse of sea.

> **The author states that it is a "June morning." Underline phrases throughout the passage that describe a spring morning.**

> **What do you think most people on the island do for work?**
> _______________________
> _______________________

ANALYZING LITERATURE

1. Main Idea How does Anne feel about this new place to which she has come, possibly to live?

____

____

2. Critical Thinking: Making Predictions Do you think Anne will stay at Green Gables? Why or why not?

____

____

ACTIVITY

Obtain a copy of *Anne of Green Gables* from a library. As you read the book, jot down some of its descriptions of Prince Edward Island in your notebook. Compare these descriptions with descriptions of regions of Canada in your textbook. Then make a list of how Prince Edward Island compares to other parts of Canada. How are they alike? Different?

Canada   Primary Source

"O Canada"
The Canadian National Anthem

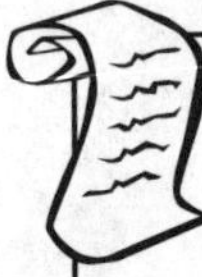

ABOUT THE READING In 1880, pianist Calixa Lavallée was asked to compose a piece of music for that year's Saint Jean-Baptiste Day celebrations in Quebec. Judge Sir Adolphe-Basile Routhier was asked to write the lyrics. After 100 years as a patriotic favorite, the resulting song, called "O Canada," was officially declared Canada's national anthem. In recognition of the country's dual cultures, there are English and French versions of the song. Both are widely sung at public events in Canada today.

As you read note general similarities and differences between the English lyrics and the French lyrics (see English translation).

English version, written by Robert Stanley Weir

The song's original lyrics are in French. The following English lyrics are based on a poem written by Robert Stanley Weir in 1908. They are not a translation of the French lyrics.

O Canada!
Our home and native land!
True patriot love in all thy sons command.

With glowing hearts we see thee rise,
The True North strong and free!

From far and wide,
O Canada, we stand on guard for thee.

God keep our land glorious and free!
O Canada, we stand on guard for thee.

O Canada, we stand on guard for thee.

Source: Canadian Heritage, Government of Canada, 2005.

WHAT DID YOU LEARN?

1. Identify What phrase in the song indicates Canada's geographic location?

____

____

2. Draw a Conclusion How do you think Canadians might feel when singing or listening to this song? Why?

____

____

French version, written by Sir Adolphe-Basile Routhier

The following lyrics are what originally accompanied Lavallée's melody. These lyrics and Robert Stanley Weir's are both still recognized and sung today.

O Canada! Terre de nos aïeux
Ton front est ceint de fleurons glorieux!

Car ton bras sait porter l'épée,
Il sait porter la croix!

Ton histoire est une épopée
Des plus brillants exploits.

Et ta valeur, de foi trempée,
Protégera nos foyers et nos droits.

Protégera nos foyers et nos droits.

English translation of the French lyrics:

O Canada! Land of our forefathers
Thy brow is wreathed with a glorious **garland** of
     flowers.
As in thy arm ready to **wield** the sword,
So also is it ready to carry the cross.
Thy history is an **epic** of the most brilliant **exploits**.

Thy **valour** steeped in faith
Will protect our homes and our rights
Will protect our homes and our rights.

VOCABULARY

garland chain of flowers or leaves

wield handle with skill

epic heroic story

exploits notable acts

valour (valor) bravery

Name _______________________________ Class _______________ Date ______________

WHAT DID YOU LEARN?

1. Explain How does Routhier's song describe Canada's land and history?

____

____

2. Make Inferences Weir's version of "O Canada" is not based on the French version. Why do you think Canadians did not translate the original French lyrics to make the official English version?

____

____

MAKE A COMPARISON

1. Draw a Conclusion Decide what you think is the theme, or main idea, of each version of "O Canada." Do they share a theme? Or does each have its own theme? Explain.

____

____

2. Compare and Contrast In the final lines of Weir's version, Canadians declare that they will guard their country. Compare and contrast this declaration with the final message in the French version.

____

____

____

Canada   Geography and History

One Country, Many Cultures

Canadian history is marked by waves of immigration to specific regions
of the country. Although people living in Canada today share a single
Canadian identity, many also identify with their culture of origin.

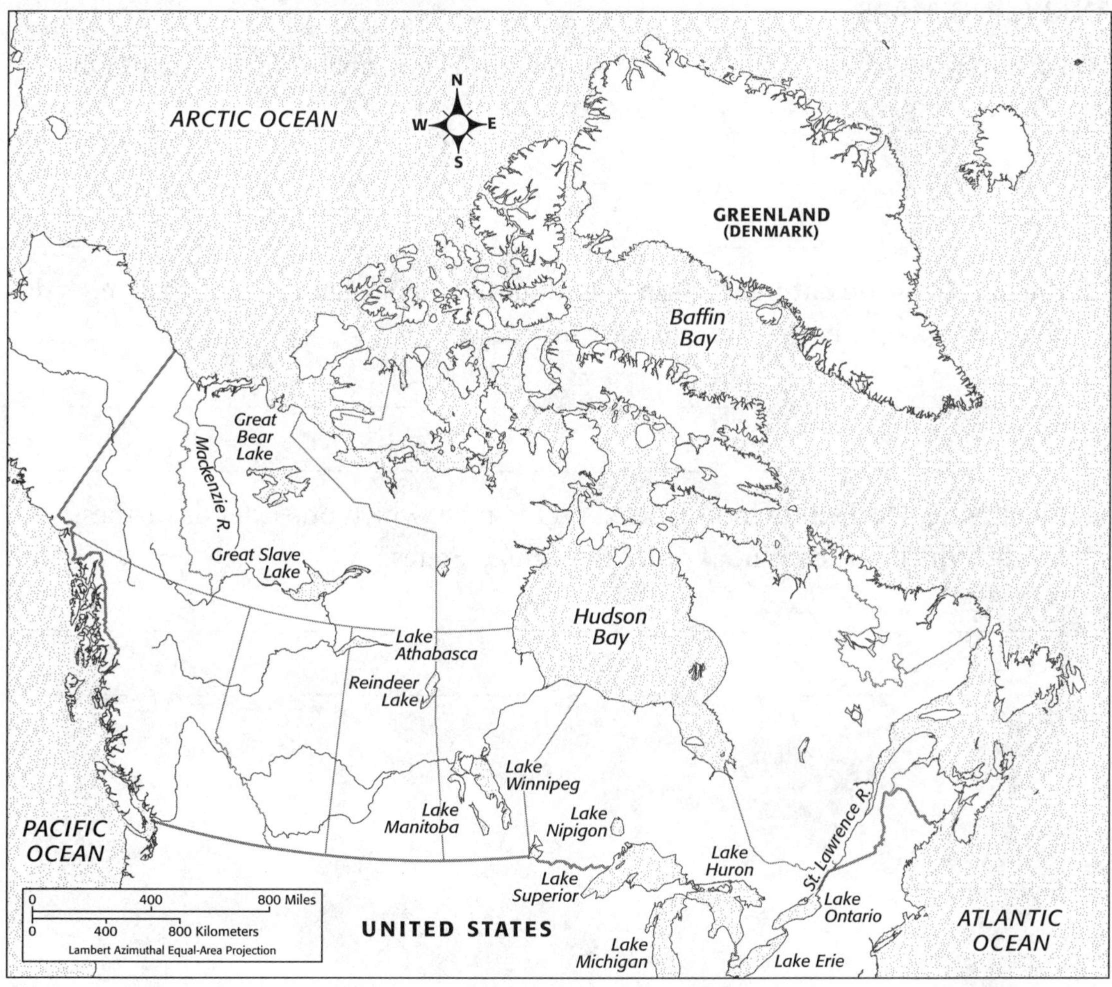

MAP ACTIVITY

1. On the map, label the following Canadian provinces: British Columbia, Quebec,
Ontario, and Nunavut. In parentheses below each name, write the region of
which the province is a part (the Eastern Provinces, the Western Provinces, the
Heartland, or the Canadian North). Use your textbook or an atlas for reference.

2. Using red, color the province originally settled by the French.

3. Using blue, color the first province to have a large Asian minority.

4. Using green, color the province that has one of the most culturally diverse cities in the world.

5. Using yellow, color the province whose people have lived there since before any immigration began.

ANALYZING MAPS

1. Location Which of the four Canadian regions is not labeled on the map? On what coast is this region located?

____

____

2. Human/Environment Interaction Based on the map, what types of challenges do you think people in the Nunavut territory face?

____

____

3. Identifying Of the provinces labeled on the map, which one extends furthest south? Which one does not border the United States?

____

____

Canada   Social Studies Skills

Using Mental Maps and Sketch Maps

LEARN

We all have an atlas in our heads. These mental maps help us get around our community—without getting lost. We create these maps in our heads, based on our experiences—when we go to school, ride to a store, or walk to a park. We can use our mental maps to create sketch maps. A sketch map can be a drawing of a place or of a route, or path, from one place to another. You can even use photos or other artwork to create a sketch map.

PRACTICE

Study this sketch map of Canada. It uses simple shapes to show the relative sizes of Canada's provinces and territories, and their relationship to each other. Which is the largest province or territory in Canada? The smallest?

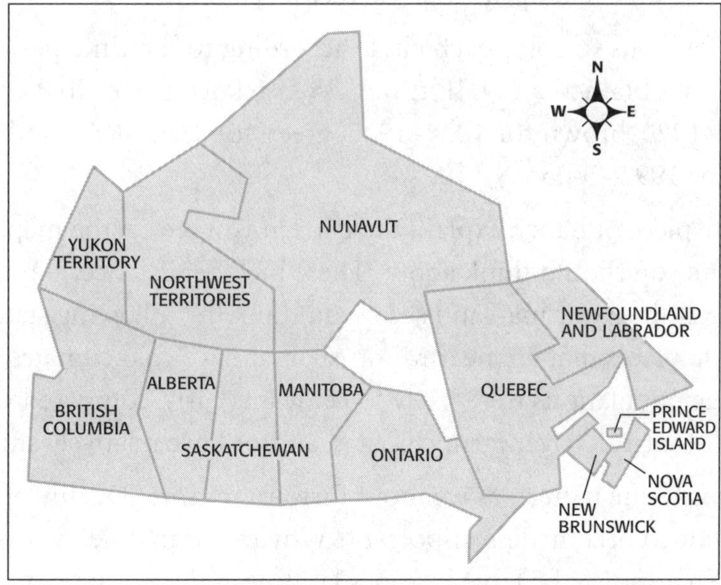

APPLY

What is your mental map of Canada? Use the map above to help you create your own sketch map of Canada on a large sheet of paper. Use shapes other than those shown here on your map. Then add drawings or photos that show something about each province and territory. For example, you may want to use colors to show different ethnic groups or symbols such as cows, oil, and fish to show resources/products. Make sure your sketch map has a legend, or key.

The Changing Geography of Ice Hockey

Ice hockey originated in northern Europe. In the 1870s, British soldiers stationed in Halifax, Nova Scotia, played on frozen ponds, and students at McGill University in Montreal played at a rink. Soon, amateur teams and leagues were organized in various Canadian cities. Toward the end of World War I, the National Hockey League (NHL) was established.

In this activity you will map the spread of the NHL, from its beginning in 1917 to 2005. In 1917, only four metropolitan areas had teams. In 2005 there were 30 teams. Follow the steps below to complete the map that follows after the table on the next page. Use it and data from the table to answer some questions about the changing geography of the NHL.

YOU ARE THE GEOGRAPHER

1. Use the list of metropolitan areas in the table on the next page and an atlas to label each circle on the map with its correct place-name.

2. Use a set of felt pens to color each circle according to the time period when that metro area first obtained an NHL team. Use black for places that got a team in the period 1917–1920, brown for 1924–1935, green for 1967–1975, red for 1976–1981, and yellow for 1992–2005.

3. On a separate piece of paper, explain the pattern you see on the map. There are various factors you should think about. These include distance, climate, and population size. Examine your map and see if you think the following statement is a fair generalization: places that are near Montreal, that have cold climates, and that have relatively large populations were likely to get teams early, compared with places that are far from Montreal, have warm climates, and/or have relatively small populations.

   What happens to the pattern as we move forward in time? Do any of the three factors mentioned become less important? Why might that be the case? To help you answer these questions, the table includes latitude and longitude for each place and its population and population rank at the approximate time it first acquired a team.

4. Canadians consider ice hockey to be their national sport and part of their national identity. Not only did Canadians start the NHL, but they continue to produce a very large number of its players. Canada exports hockey players not only to the United States but to other places where hockey is growing in popularity, including Europe and Japan.

   How might you feel about the changing geography of ice hockey if you were a Canadian? Would you mourn the loss of NHL teams and players or celebrate the diffusion of your sport to the rest of the world?

The Changing Geography of Ice Hockey, *continued*   Geography for Life

Metropolitan Areas That Have (or Once Had) NHL Teams

Year acquired	Metropolitan Area	Latitude & Longitude	Population at the time	Rank at that time	Current Team(s)
1917	Montreal	45.30N, 73.35W	1,086,000	11	Canadiens
1917	Toronto	43.40N, 79.23W	901,000	14	Maple Leafs
1917	Quebec	46.49N, 71.13W	207,000	54	None
1917	Ottawa	45.25N, 75.43W	197,000	58	Senators
1920	Hamilton	43.15N, 79.52W	190,000	62	None
1924	Boston	42.15N, 71.07W	2,308,000	5	Bruins
1925	Pittsburgh	40.26N, 80.01W	1,954,000	7	Penguins
1925	New York-Northern NJ	40.40N, 73.58W	10,901,000	1	Rangers, Islanders, Devils
1926	Chicago	41.49N, 87.37W	4,365,000	2	Blackhawks
1926	Detroit	42.22N, 83.10W	2,105,000	6	Red Wings
1930	Philadelphia	40.00N, 75.13W	2,847,000	3	Flyers
1934	St. Louis	38.39N, 90.15W	1,294,000	8	Blues
1967	LA-Anaheim-Riverside	34.03N, 118.14W	7,176,000	2	Kings, Mighty Ducks
1967	SF-Oakland-San Jose	37.48N, 122.16W	4,399,000	6	Sharks
1967	Mpls.-St. Paul	44.58N, 93.15W	2,037,000	16	Minnesota Wild
1970	Buffalo	42.54N, 78.51W	1,590,000	29	Sabres
1970	Vancouver	49.76N, 123.06W	1,268,000	31	Canucks
1972	Atlanta	33.45N, 84.23W	1,832,000	19	Thrashers
1974	Wash., D.C.	38.50N, 77.00W	2,815,000	9	Capitals
1974	Kansas City	39.05N, 94.35W	1,293,000	30	None
1976	Cleveland	41.30N, 81.42W	1,950,000	18	None
1976	Denver	39.44N, 104.59W	1,464,000	23	Avalanche
1979	Hartford	41.45N, 72.40W	1,052,000	38	None
1979	Edmonton	53.33N, 113.28W	741,000	54	Oilers
1979	Winnipeg	49.53N, 97.09W	592,000	68	None
1980	Calgary	51.03N, 114.05W	626,000	61	Flames
1992	Tampa-St. Petersburg	27.57N, 82.25W	2,068,000	23	Lightning
1993	Miami-Ft. Laud.	25.45N, 80.11W	3,193,000	12	Florida Panthers
1993	Dallas	32.45N, 96.48W	3,885,000	9	Stars
1996	Phoenix	33.30N, 112.00W	2,111,000	22	Coyotes
1997	Raleigh-Durham	35.45N, 78.39W	735,000	61	Carolina Hurricanes
1998	Nashville	36.10N, 86.48W	985,000	42	Predators
2000	Columbus	40.00N, 83.00W	1,377,000	32	Blue Jackets

The Changing Geography of Ice Hockey, *continued*

Geography for Life

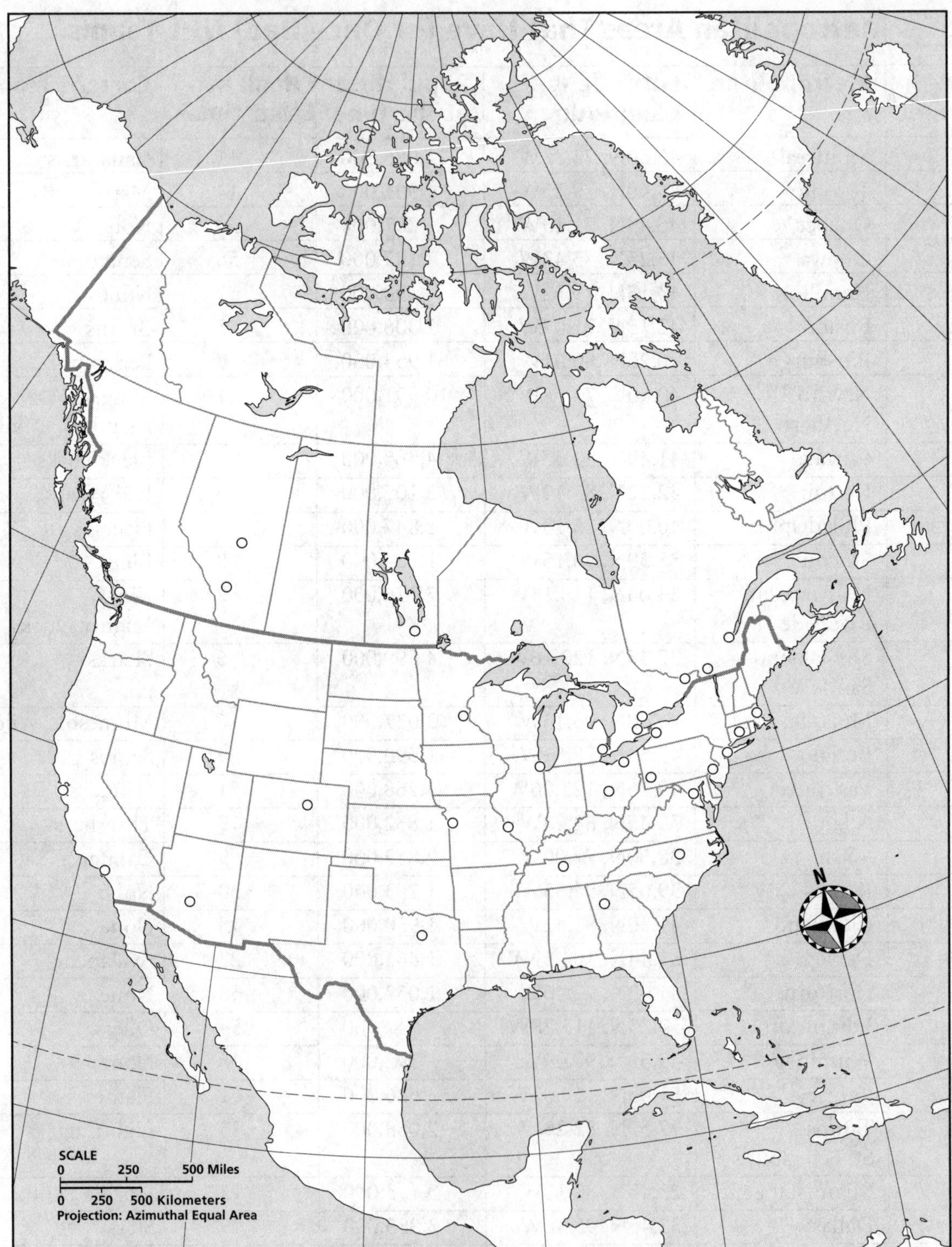

Canada

Can Canada Survive?

In 1995 the province of Quebec held a referendum on independence. By a narrow margin—one percentage point—the people of Quebec decided to remain part of Canada and not to form an independent nation. However, 60 percent of French-speaking Quebecers voted for independence, and its supporters promise to put the subject to another vote in the near future. To learn more about Canada's crisis, study the information below and answer the questions that follow.

Largest of Canada's ten provinces, Quebec is the only predominantly French-speaking political entity in North America. . . . In recent years the province has been the focus of the nation's worst unity crisis since its birth in 1867. . . .

That the Canadian federation has survived this long has to do with various compromises made on behalf of Quebec. The province retains its own distinct legal system and has been given special powers of immigration, enabling it to attract French-speaking newcomers. In the past 20 years, however, these compromises have worn thin, and constitutional amendments to patch thing up have failed to win approval. . . .

[In Montreal, Quebec's largest city and home to nearly half its people,] 58 percent of Montrealers voted against independence in 1995. . . . One in five Montrealers [is] out of work, 22 percent [live] below the poverty line—the highest rate of any city in Canada. . . . Companies are either not investing or simply packing up and moving out of the province. . . . For many Montrealers . . . the solution has been to leave. More than 325,000 people have moved out since 1971; often it is the younger generation of English-speaking Quebecers, those unencumbered [unburdened] by houses and children, who are voting with their feet.

Since the 1995 referendum, it has become clear just how close Canada came to splitting apart. Jacques Parizeau, a hard-line sovereigntist [separatist], was premier of Quebec during the election campaign. [In May 1997] he revealed in a memoir that he had received secret assurances that the French government would quickly recognize a sovereign Quebec and would influence other francophone [French-speaking] nations to follow suit. . . . The province's minister of finance had quietly set aside

From "Quebec's Quandary" (retitled "Can Canada Survive?") by Ian Darrach from *National Geographic*, May 1997. Copyright © 1997 by **National Geographic Society**. Reprinted by permission of the publisher.

a reserve of 27 billion dollars to buy Quebec bonds in case they were dumped by panicky investors. A week before the referendum, a letter was sent to Quebec-born soldiers in the Canadian Army urging them to transfer their loyalties to a new Quebec army in the event of a "Yes" vote. . . .

Now the debate has shifted to setting the practical ground rules for independence: What majority would be required in the next referendum? Would a unilateral [one-sided] declaration of independence violate the Canadian constitution? The federal government has referred that question to the Supreme Court of Canada. [In 1999 the Supreme Court ruled that a simple majority (50 percent plus one vote) was not a large enough margin to allow secession. But the Court did not say how large a majority was needed. A number of bills have been introduced into Canada's Parliament to establish that figure.]

What if Quebec was to ignore the constitution and the Supreme Court? Jean Chrétien, [former] Prime Minister of Canada and a fellow Quebecer, says that if a clear majority choose independence in a fair vote, he is not prepared to launch a civil war to preserve the nation.

1. Use a physical map of Canada from an atlas or another resource to describe how Quebec's independence would affect the geography of Canada.

____

2. Why might the St. Lawrence River become an issue if Quebec gains independence?

____

3. What connection exists between Montreal's problems and the independence movement?

____

4. How large a majority should support Quebec's independence for it to become a reality? Should most English-speaking as well as French-speaking Quebecers have to support it? What about the wishes of other Canadians in this matter? On a separate sheet of paper, describe a system that you think would be a fair way to settle the question of Quebec independence. Explain why your system is the proper way to decide this issue.

Canada   Focus on Reading

Understanding Lists

Writing lists can help you better understand your reading. Lists can also
help you organize your studying and writing. Make a list for the chapter
on Canada. Follow the steps below to write your list.

SETTING UP YOUR LIST

Write "Canada" as the main heading on the chart below. Then label
the first column "Physical Geography." Label the second "History and
Culture." Label the third "Canada Today."

WRITING YOUR LIST

1. Add notes about Canada's physical geography in Section 1 to your list.

2. Add notes about Canada's history and culture from Section 2 to your list.

3. Add notes about Canada today from Section 3 to your list.

If you need more room for writing, use another sheet of paper.

USING YOUR LIST

Look over your list. Check that it is complete. Are all the key facts about
Canada on your list? Use your list to study for a test or quiz on Canada. You
can also use it to help you create your Radio Ad for Canada (see next page).

Canada

Focus on Speaking

Creating a Radio Ad

The Canadian tourism board needs you to create a radio ad to attract
visitors to Canada. Your ad can only be one minute long. You need a
script for your ad. Follow the steps below to write your script. Then
present your ad to your class.

PREWRITING

Writing About Canada On a separate sheet of paper,

- make a list of information about Canada's physical features, climate, and
  resources from Section 1.

- add details about the history and culture of Canada from Section 2.

- describe details about Canada today from Section 3.

WRITING

1. Writing Your Radio Script On a
separate sheet of paper, make a script
model like the one shown here. Use
your list(s) to write your radio script.
Remember, your job is to convince
people to visit Canada. Make sure you
use descriptive and persuasive language
in your script. Use short, snappy, ear-
catching phrases. Describe Canada in
a way that's sure to get your audience's
attention!

Introducing Canada: _________________   ___________________________________   **Canada's Amazing Physical Geography:** _   ___________________________________   **Canada's Rich History and Culture:** ______   ___________________________________   **Canada Today Is Spectacular:** _________   ___________________________________   **Canada—It's Worth the Trip:** __________   ___________________________________

- Write an introduction that sets the
  tone for your ad. Use adjectives that capture the essence of Canada as a whole.

- Add information about Canada's most important physical features, climate,
  and resources. Make sure you focus on what would attract people to Canada.

- Next write about Canada's history and culture. Again, focus on what people
  would like most.

- Then describe Canada today. Use descriptive language so Canada comes alive
  for your listeners.

- Finally write a conclusion. It should remind listeners of what's interesting
  about Canada as a country—and convince them to make the trip!

EVALUATING AND PROOFREADING

2. Evaluating Your Radio Script Use the questions below to evaluate and revise your script.

Rubric

- Can you read your script aloud at a steady speed in one minute?

- Does your introduction build interest in Canada and lead into the ad?

- Are your ideas about Canada's physical geography, history and culture, and Canada today clearly stated?

- Is your conclusion snappy and leave your audience interested in knowing about Canada?

- Have you practiced your radio ad enough so that you can use your voice, tone, and rate of speech to create interest in visiting Canada?

3. Proofreading Your Radio Script Before you present your ad, check the following:

- Punctuation, grammar, capitalization, and spelling

- Topic sentences and vivid supporting details for each paragraph

PRESENTING YOUR RADIO AD

4. Present Your Ad to Your Class Think about which radio ads capture your attention. Decide what makes those radio ads successful. Is it the tone of voice? The excitement of the speaker? Try using some of the presentation techniques as you present your radio ad. Use the following tips to help you:

- Speak clearly and with enthusiasm. Make Canada come alive for your classmates.

- Stay within your time limit. A radio ad should be short and leave the listener wanting to know more.

- Grab your listeners' attention with your tone of voice. Change your rate of speech to generate interest.

- Try not to pause or say "um." This will take up time and make your audience think the topic isn't interesting.

- Have fun! Your attitude will build interest in Canada.

Canada

Chapter Review

BIG IDEAS

1. Canada is a huge country with a northerly location, cold climates, and rich resources.
2. Canada's history and culture reflect Native Canadian and European settlement, immigration, and migration to cities.
3. Canada's democratic government oversees the country's regions and economy.

REVIEWING VOCABULARY, TERMS, AND PLACES

Choose from the three letter sets on the right to complete the vocabulary terms.

1. NEW___ ___ ___ INT	**NDR**
2. CANADIAN S ___ ___ ___ LD	**LIS**
3. TU___ ___ ___A	**ONT**
4. P ___ ___ ___ INCE	**EBE**
5. QU ___ ___ ___C	**HIE**
6. ___ ___ ___ARIO	**SPR**
7. REGIONA___ ___ ___M	**IME**
8. MARIT___ ___ ___	**ROV**

COMPREHENSION AND CRITICAL THINKING

Read each of the following pairs of sentences, and cross out the **FALSE** sentence.

1. **a.** Russia is the only country in the world that is larger than Canada.
   b. The United States is the only country in the world that is larger than Canada.

2. **a.** The climate is generally the same throughout Canada.
   b. Canada's climate ranges from mild to freezing, depending on region.

---

3. a. The first Europeans in Canada were the French.
   b. The first Europeans in Canada were the Vikings.

4. a. Canada's economy suffered in the early 1900s due to a wave of immigration.
   b. Immigration helped Canada's economy to boom in the early 1900s.

5. a. Canada's democratic government is led by a prime minister.
   b. Canada's democratic government is led by a president.

6. a. English is the major language in Quebec.
   b. French is the major language in Quebec.

7. a. Canada's leading trading partner is China.
   b. Canada's leading trading partner is the United States.

REVIEWING THEMES

In the space provided, explain how each term relates to the theme listed below.

Theme: *human-environment interaction*

1. pulp ____

____

2. Montreal ____

____

REVIEW ACTIVITY: CANADA QUIZ GAME

Work with a partner or small group to create your own Canada game. Base your game on what you have learned about Canada. Create 40 cards, and on each card write interesting (and tricky) questions based on Canada's provinces, territories, cities, and physical features. On the other side of each card write answers. Write a set of rules for your game. For example, how many people can play and how a person (or team) can win. If possible, play the game with your classmates. Here are a few ideas to help you get started.

Alberta	Charlottetown	Edmonton
Montreal	Nunavut	Niagara Falls
Ontario	Ottawa	Quebec
Rocky Mountains	St. Lawrence River	Vancouver

Vocabulary Builder

SECTION 1
1. St. Lawrence River
2. Pulp
3. Rocky Mountains
4. uplands
5. Niagara Falls

Students' summaries will vary but should reflect an understanding of the section's terms.

SECTION 2
1. F; Toronto
2. F; provinces
3. T
4. F; Quebec
5. T

6. Pacific coast; large Asian minority; mountainous
7. early French settlement; now Quebec; seized by England
8. linked Canadian coasts; built in 1880s; transcontinental
9. neighborhood; unique place; people speaking Chinese
10. Indians and Inuits; bison hunters and seal hunters; interior plains and far north

SECTION 3
1. d
2. b
3. a
4. b
5. d

Students' poems and stories will vary.

Biography

DOUGLAS JUNG

WHAT DID YOU LEARN?
1. Jung was among the first Chinese-Canadians to enter the Canadian Army. He fought in World War II.
2. possible answer—Jung's greatest achievement was being elected the first Chinese-Canadian Member of Parliament.

Activity responses will vary. Key Canadians of Jung's time include John Diefenbaker, John Peters Humphreys, and Yousuf Karsh.

Biography

MARY PICKFORD

WHAT DID YOU LEARN?
1. She wanted to succeed on Broadway, partly to help her family financially.
2. She had become the most well-known female star in American film.

Activity responses will vary. Her movies include *Cinderella* (1914), *Little Lord Fauntleroy* (1921), and *The Taming of the Shrew* (1929).

Literature

CALL-OUT BOXES
1. *delightful thrill* and *horrible remembrance*
2. *cheery sunshine, blue sky, cherry-tree in full bloom, and sweet fragrance drifted up to the window on the morning wind*
3. Students may say fruit picking or farming.

ANALYZING LITERATURE
1. possible answer—Anne thinks this new place is beautiful. She hopes she can stay.
2. possible answer—Anne will stay at Green Gables, as the title implies.

Primary Source

WHAT DID YOU LEARN?
English version
1. The True North
2. possible answers—proud, enthusiastic, sentimental, patriotic; tells about the value of the country and its people

French version
1. Its land is wreathed with glorious flowers; its history is a series of heroic events.
2. possible answer—English-speaking Canadians wanted a version meant to be sung in English; translated into English, the French version might not rhyme or work well with the melody.

MAKE A COMPARISON
1. possible answer—The versions share a main theme of Canadian pride. The English version is also about freedom; the French version is about historic celebration.
2. possible answer—Both versions suggest that its citizens will defend Canada. The main difference is in the language used.

Geography and History

MAP ACTIVITY

1. British Columbia: Western Provinces; Quebec: Heartland; Ontario: Heartland; Nunavut: Canadian North
2. Quebec should be red.
3. British Columbia should be blue.
4. Ontario should be green.
5. Nunavut should be yellow.

ANALYZING MAPS

1. Eastern Provinces; Atlantic
2. possible answer—keeping warm, growing food, accessing goods and services
3. Ontario; Nunavut

Social Studies Skills

Practice: Nunavut; Prince Edward Island
Apply: Sketch maps will vary, but should include the relative sizes of each province and territory, and a distinct aspect of each.

Geography for Life

1. Labels should match the table listings.
2. Colors should match the table listings.
3. Responses will vary, but should note that more teams are in the U.S. than in Canada.
4. possible answer: For Canadian cities that lost teams, fans may no longer be as interested in following the NHL.

Critical Thinking

1. possible answer—It would reduce Canada's size and separate eastern provinces from other provinces/territories.
2. possible answer—Who controls the river may lead to a conflict.
3. possible answer—The problems facing Montreal lead many to support independence.
4. possible answer—If over two thirds of people in Quebec vote for independence, it should be allowed—subject to approval by a nationwide referendum. This allows all Canadians to decide Canada's future.

Focus on Reading

UNDERSTANDING LISTS

Details on lists may vary, but should include key facts from all three sections.

Focus on Speaking

CREATING A RADIO AD

Scripts for ads will vary, but should use persuasive and descriptive language about key facts on Canada; have a strong beginning, middle, and end; and be able to be read in about one minute. Students should speak clearly, without significant pauses, and with enthusiasm. The tone and rate of speech should vary to help increase audience interest in the topic.

Chapter Review

REVIEWING VOCABULARY, TERMS, AND PLACES

1. SPR
2. HIE
3. NDR
4. ROV
5. EBE
6. ONT
7. LIS
8. IME

COMPREHENSION AND CRITICAL THINKING

1. b
2. a
3. a
4. a
5. b
6. a
7. a

REVIEWING THEMES

1. Trees are used to make pulp, which in turn is used to make newsprint.
2. Montreal has very cold winters, so people have built underground passages and overhead tunnels to move between buildings.

REVIEW ACTIVITY: CANADA QUIZ GAME

Students may need to review Sections 1 and 3, to be sure their games include questions about Canada's provinces, territories, cities, and physical features. Higher-quality games will cover a greater variety of places, a theme to their choices of places, and a consistent rationale for how the game is played.